FIELD NOTES ON QUEEN REARING

Oliver S Field

Jointly published by:
The International Bee Research Association,
a Company Limited by Guarantee, 1, Agincourt Street, Monmouth, NP25 3DZ (UK) &
Northern Bee Books, Scout Bottom Farm, Mytholmroyd, Hebden Bridge HX7 SJS (UK).

Obtainable from:
www.ibra.org.uk & www.northernbeebooks.co.uk

IBRA is a not-for-profit organization that exists to increase people's awareness of the vital role of bees in agriculture and the natural environment. It aims to promote the study and conservation of bees, and provide useful and reliable information to beekeepers and bee scientists all over the world.

Reprinted: 2022

British Library Cataloguing-in-Publication Data
A cataloguing record for this book is available from the British Library

Written by Oliver S Field
Edited by Richard Jones
Complied by Tony Gruba

IBRA Proof Editor - Stuart A. Roberts

ISBN 978-1-913811-13-6

FIELD NOTES ON QUEEN REARING

Oliver S Field

Reprinted from a talk originally given to the
Scottish Beekeepers Association
12 September 1998
Updated with revisions 2008

CONTENTS

The Author

Oliver Field has been involved with honey bees since boyhood, first keeping a few hives in his early twenties and then being kept by bees in his early thirties, by which time he was managing almost a thousand hives at Chiltern Honey Farm in Oxfordshire. He then went on to set up his own business.

Now retired but retaining a practical interest in Field Honey Farms which are now managed by his son, Robert. Currently, they have 500 colonies with a potential crop of 25 tons of honey per season.

He has written two books on the subject of bees, the best known of which is *"Honey by the Ton"*.

He has been especially interested in Queen-rearing since the early '70's and today he and his son produce all their own queens at Purbeck in Dorset.

Oliver has also been involved in a number of overseas projects in Africa and the Middle East, advising on honey production and providing expertise in moveable frame hives.

FIELD NOTES ON QUEEN REARING

It is now more than thirty years since I turned to bee farming for financial support and not without some difficulty I found it! At that time I bought two hundred hives of bees with very little knowledge on how I was going to manage them.

I had met Tony Rowse, through my brother who was friendly with Tony's son, who referred me in turn to his brother David. He considered that David knew rather more about honey farming than most other mortals. I made an appointment and met David in his queen rearing apiary where he was checking through some fifty double nucs to see which queens had mated and which had not. I peered over his shoulder and we looked for eggs. He held a frame up to the light and paused for a moment, " Well my boy," he queried with a smile, " What are you going to do about queens?"

I thought for a moment, "I am not sure," I took a deep breath, "I haven't really given it a thought."

"In that case you had better start thinking right now, you either buy in or you breed your own, but good queens are the basis to all honey production, you will never produce much honey with poor queens you know."

That is how I learnt my first major lesson in honey farming and I have been trying to find the right answer ever since! Good stock is so important and yet many beekeepers and even honey farmers pay scant attention to what they breed from. In many cases when a stock is found to have swarmed, a cell is left to perpetuate the tendency into the future. Then in the next season the same thing happens again, with the honey crop being cut back year after year and swarming forever on the increase.

To start with, in today's climate, where do we go for good stock? In the past it was easier, the leading bee farmers were breeding good quality bees from home reared strains that went back years. Then, soon after I took up honey production, everybody started importing queens from all over the globe. There were American queens, Australian queens, New Zealand queens, queens from Israel, Turkey, Romania and even Hawaii. Each producer claiming that his strain was the best and then selling them all round the country. Very soon our useful old homebred bees became diluted with this new influx of genes.

Like most young men, in the early days I was tempted and I tried several of these foreign strains. I soon found that bees from other climes certainly did not like our weather and by the middle seventies I had completely given up all foreign stock.

For myself I can safely say that queens bred from bees that have evolved in your own area will always be the best. I can remember David Rowse remarking that the Carniolan (*Apis mellifera carnica*) queens that he imported from Italy were very fine queens, but that the first cross with his own drones were always better. I once tried six of these crosses and they were quite excellent.

From the above you can see that if I were going to start to raise queens today, I would look for a good queen in a local hive. Something with a steady track record, which could be traced back for at least five years.

THE BREEDER

Having found the stock from which we wish to breed, the next quesion to ask is, *"What particular traits do we wish to see in our next generation of bees?"*

Point 1 for me will always be supersedure. From the very beginning in my early days working with Harry Wickens, he impressed on me the importance of this trait. It was a priority from which he would not be deflected. We would check the notes of the whole bee farm and select the best from a number of superseding strains, noting other important traits as we went along. Swarming is the curse of bee farming and the less of it you see the better your life will be. If you spend many years breeding from superseding strains, as I have done, you will find that swarming decreases year on year.

At Chiltern Honey Farm, during the years I was with them, swarming was at thirty to thirty-five percent per annum. After I had left them and I was on my own, I reduced the swarming rate in my own apiaries to nearer twenty percent, though I do feel that it has crept up again in recent years.   A year or two ago I spoke to an Irish bee farmer and he told me that he had a swarming level of eighty seven percent, to me such a level would make the chance of a profit very slim indeed! Swarms of bees can be collected right through the summer and hives can be filled with them, without much effort, but that's where it ends, controlling them and getting a crop of honey is more than a lifetime's work!

The more superseding strains are encouraged, the less the workload is at the busiest time of the year. Superseding type queens will mother superseding type drones and as your breeding gets under way and you re-queen more stocks in the out apiaries, the trait will spread.   At a level of thirty to thirty-five percent, you can cope with the swarming, it will be enough to build up your losses, yet not too much to really damage your honey crop. Over fifty percent and you will start to lose honey in a big way and get thoroughly overworked and frustrated into the bargain.

Point 2 on my list of traits must be the ability to get honey. Many foreign strains just have not got the guts to find a crop in our changeable weather. In selecting a breeder there is a need to look at her past honey yields and, if you can, those of her mother as well. At Field Honey Farms our hive average for the last ten years has been something over a 100 lbs (45 kgs) of honey from every hive managed. This has been achieved on a bee farm of 300 - 330 stocks. Long before you race to four, five or even six hundred stocks, take a long hard look at your averages, the answer too often is not more bees, but better queens. Remember every pound more honey produced from the same base, will be a profit, but on the other hand every pound produced from a new stock will carry a cost. The larger that base is the greater the cost, there is a balance you know!

Point 3 on my checklist must be aggression. Aggressive bees are not amusing things to keep or handle. Over the last thirty or so years I have, on occasions, had to cope with aggression and I have the mental scars to remind me! In my early days at Chiltern Honey Farm I remember one stock that got totally out of hand and I had to admit defeat. I struggled back to Harry Wickens covered in bees and asked for his help. I can still see and remember that grin through the veil, as he stoked up his smoker and set about the offending stock. Such things must be avoided and all such hives should be given a new queen at the earliest opportunity. Never breed from such stock: any queen cells found in such a colony must be shown no mercy! On the other hand there is a certain amount of merit in considering what we call hybrid vigour!

My son Robert bred some of the best honey gathering bees that we have had at Field Honey Farm in recent years. He crossed a light queen with some of our darker drones. The bees would defend their home to the last and needed plenty of smoke to control them with the result that, when trying to find a queen for clipping, we would almost have to draw lots as to who was going to take his gloves off and catch her! Even so these bees all produced huge crops of honey and not too many of them swarmed. The moral must be, "try the cross, but don't continue the line and re-queen if necessary". You just may be lucky.

Point 4 This is the thorny question of resistance to disease. All bees can get sick, of that there is no doubt, but I have found that bees that come from foreign climes tend to get sicker faster and more often than bees bred at home from home stock. Some time ago when considering going to West Africa to look at bees on a village project, I was appalled at what would have to be pumped into me to keep me alive. I rest my case, if I cannot live in West Africa without help, it is hardly surprising that bees from other parts of the world need help to survive in our fickle climate. Suddenly they are faced with diseases and parasites they have never met before, little surprise that they succumb at the sight of an Acarine mite peeping through the hive entrance, let alone a Varroa mite.

The more we breed bees that are used to our multitude of problems and the less we breed from foreign strains, the better off we will all be, of that I am sure. To an extent the problem will answer itself, for if we have already selected a queen that has given a large crop of honey for several years, this will indicate that she is well used to and has overcome at lest some of the health problems that our bees face today.

Point 5 Longevity in queens is a trait that is well worth watching, I have once had a superseding home bred queen live for five seasons and be superseded herself in her sixth summer, having never once tried to swarm. Just think what a trouble free life you might have with a couple of hundred similar mothers. In the days of Chiltern Honey Farm, Harry Wickens always said that it was best to re-queen a honey stock every two seasons. On the other hand, many were the occasions when he would look at a three-year-old queen about to go into the winter, and say, "She looks fine to me, I think we will give her another chance." Often it was a mistake and we would find a new young queen in the spring. Autumn superseding would have moved her on, but as can be seen this was no loss either. Both ways the bee farm would win.

I would like to think that if the five above mentioned factors are followed then, once breeding takes place, the result will be fine active honey producing queens with a good long shelf life.

When considering a breeding programme the clipping of the wing or wings of every mated queen is vital from a recording point of view.   Once a queen has been laying fertile eggs and a patch of sealed worker brood seen, she should be clipped and used to re-queen a full sized stock. If this is done there can never be any doubt as to her age. Then, as soon as a queen is seen in a hive with her wings intact, the beekeeper knows there has been a change.   It will of course help with swarm control into the bargain, as you will lose fewer bees when the going gets tough in May and June.

Once we have selected our new breeder queen for the coming season, she must be looked after.   There is no point in leaving her in a honey stock to exhaust herself with egg production. So make up a four-comb stock from her hive and take her home with it. Put it in a six-comb nuc box, add a good comb of sealed food and place a sheet of foundation in the middle. The foundation is there, so that once she has settled down and got back on to the lay, you will have eggs and small grubs on a freshly drawn comb. This makes life far easier when you come to graft a batch of grubs to produce your new queens.

If the weather is inclement, this little stock should be given a small feed of dilute sugar syrup, for we do not want any question of it getting short of food.   If, in the next week or two, the stock seems to be over breeding take out a comb of sealed brood and replace with another sheet of foundation. In this way you will keep your grafting material available on fresh foundation, and stop the stock from getting tight for room.

A GRAFTING SHED

Over the years, like David Rowse and Harry Wickens, I have always had a grafting shed on the site of the breeding apiary. Small sheds can easily be bought, erected and, during the winter, can be used for storage, being filled with equipment, mini nucs, drawn combs and the like. In recent years I have fixed the grafting shed on to six short posts set eighteen inches clear of the ground. This method will make it very difficult for mice or even rats to find their way in and will protect your stock from damage. Nothing is more annoying than to find all your little mini nuc frames carefully drawn the year before, torn to shreds in the winter, or eaten by wax moth. Treat the shed in the autumn for moth. Secondly there will be a draft space under the shed, so keeping the floor dry and sound. A window, facing west is ideal and a flap type table under the window can be hinged, so as to create a grafting platform when needed. The roof, of course, must be rainproof and the roofing felt well tacked on.

I have in the past used a green house as a grafting shed and this will do, but on very hot days it may get too hot and the little grubs can get over heated and dried out. If the soil is well doused with water just before you graft this problem can be controlled. In places like Australia the heat can be a real problem and grafting is often carried out in the open air and dry grafting at that. We on the other hand will usually suffer from a condition of chill to cold when we wish to graft so a building of some sort is needed.

GRAFTING

At this point we can consider raising some queens. We have a breeder queen in a six comb stock at the side of our well appointed grafting shed. We will now need a cell builder, that is, a stock without a queen but with plenty of bees and young brood. Make very sure there is not a virgin with it, or a nasty little queen cell tucked away in some corner of one of the frames, if there is, then all your work may count for nought!

Make up your cell builder from several stocks of bees from at least three different hives. This will confuse the bees and they won't fight, they don't know whom to attack! They will also be much more likely to accept batches of grafted grubs; in fact they will be delighted with them.

Grafting is a very delicate operation and needs a great deal of care for it to be successful. The size of the little grub is vital and should be twenty four to thirty six hours old. Once you start to try to graft, you will realise the great advantage of having that beautiful freshly drawn comb to work on, you will be able to see what you are doing. You will see the eggs on the outside of the brood and as your eye wanders inward towards the centre of the comb, you will see the freshly hatched grubs and then the larger grubs, until you may find some sealed brood in the centre patch.

Look for the most recently hatched grubs and try to move the smallest ones you possibly can. For this purpose you will need the finest little grafting tool you can make. Do not end up trying to move a chicken's egg with a garden shovel, or you are likely to break it. The same is true with very small delicate grubs. If you end up chasing them round their cell with a huge piece of bent metal you will probably damage them and they will not turn into good queens.

The art is to slip the little spoon-like tip of the tool under the grub and gently lift, if you don't get it first time, try another grub. If you fold down the sides of the cells of a row of the right aged grubs, you can then work along the batch and leave those you think you might have damaged. It is far better to leave grubs behind, than find you have only achieved three queen cells out of twenty attempts, the rest having been licked out of the cups by the bees.

Queen cups can be made at home with a former, but to day it is possible to buy artificial cups that you can line with your own wax cup, or use as they are. For myself, I would always like to graft into a true beeswax queen cup that I have made myself, but then as my son will tell you, I am old fashioned!

Again, I like to use a drop of royal jelly in the base of the queen cup to float the little grub into. I also think that it is a good idea if the royal jelly is warmed slightly with the addition of 50 percent warm water. This mixture can then be placed in the base of the cell and the little grub will not freeze to death before she gets back to the cell builder colony. It may take several minutes to graft a bar of a dozen cups, especially if you are new to it, and if it is a cold day, you could be in trouble grafting dry.

Once the bar of queen cups has been grafted, and there is a tiny grub glistening in each one, it should then be placed in the centre of the cell builder. A space for the bar should already be there, you do not want to have to start fiddling around with frames and a hive tool and let those precious grubs chill off.

ILLUSTRATIONS

The mating yard at Chiltern Honey Farm. On this site there were 40 double nucs on standard frames working from separate hive ends. The aim was to get 200-250 mated queens from this unit each season. Note the telegraph pole sections that were used for stands.

A load of over 70 hives returning from pollination to be redistributed to their summer sites. The picture shows Paul Deakin and the author hard at work.

A typical commercial shed that would be ideal for grafting operations. It is in the shed that the serious manipulation of the breeder hive takes place. This means that the combs containing the developing brood do not die due to change in temperature. The shed is heated prior to any breeding activity.

A comb with developing brood is taken from the breeding hive to the grafting shed.

Each grub is placed in an artificial queen cell and has the potential to become a virgin queen.

The author placing a graft of cells in the cup stage into a queenless stock. This hive then develops the cells in an attempt to generate a replacement queen. Such a hive is, therefore, termed a cell builder.

After a few days the bar containing the artificial
cells is removed from the cell builder to see
just how many have been accepted.

A successful graft would be in the region of 70-80% -
12 good cells on a bar of 16.

Robert Field's insulated box for
trans-porting queen cells and
maintaining temperature so that
larvae are not chilled.

A successful queen cell is inserted into the mini nuc, which is a box
containing a few hundred bees shaken in a few days previously.

Each stand of 4 boxes has a different coloured roof and all the nuc boxes have a different coloured entrance, everyone of which is facing a different direction. This is to enable the virgin queens to recognise their own hive as they arrive back from their mating flight.

MINI NUCS

At this stage while we are waiting for our superb young queens to develop in their cells, we must turn our minds to mini nucs. It is no good producing fifty, a dozen or even five beautifully formed and about to hatch cells, if you have nowhere to put them.

When I was at Chiltern Honey Farm, and also learning my queen rearing from David Rowse, both honey farms used four-comb standard nucs in a double box. That is to say, eight standard combs separated down the middle, with entrances at opposite ends. These units could be united if the central board was removed and then they could be over wintered as a full eight-comb stock of bees. In the spring they were split up again, when queen rearing started. Each half was given a queen cell, while the old over wintered queen could be used to make up a new hive.

When I left Chiltern Honey Farm and went back on my own I gave this matter a lot of thought I decided that if I made up some 9 inch (230 mm) square nuc boxes, 6½ inches (165 mm) deep, I could place four such boxes in a square on top of a Modified Dadant (MD) brood chamber. In this way it would be like putting on a super in the spring. I had one hundred such boxes made, each with its own little crown board, and my dutiful son worked very hard one winter making up all the little frames. He made them by cutting down ordinary MD super frames to 6½ inches (165 mm) and adding a new end to each. They were then wired, given foundation and were ready to be placed on to a strong MD brood nest in the coming spring. Hopefully at the time of a rape honey flow!

We planned that the bees would draw out these little boxes, partially fill the frames with rape nectar and would be all ready for use. Then, at the same time, with luck, we would have the cells finished in the cell builder. Of course nothing ever goes to plan, a cold spring and they would not draw out our little boxes, a hot spring and before you know it the whole lot is sealed solid with rape honey. That said, it was a fair success. We got our little frames drawn and cleared off boxes, honey, and bees in one swoop. They were then taken back to the breeding apiary held together in a MD roof covered with a crown board. Each little nuc had a tiny 1 inch (25 mm) entrance, which was placed against the side of the next nuc, so the bees could not get out.

Once home they were placed on a flat board nailed to the top of a length of telegraph pole set in the ground giving a convenient working height. The bees were kept shut in for two to three days, to stop them wandering and then they were turned with an open entrance on each of the four sides of the square. The little crown boards were put in place and a normal MD roof covered the lot. We now had four mini nucs in a square, with bees flying from each side. At this stage each block of four would be ready for the introduction of the sealed queen cells, one cell being tucked into the middle of each nuc a day or two before it was going to hatch.

There was a tendency for one nuc on the block to suck bees from the others and although they all started at the same strength, it never lasted long. Even so over the years it has worked well and we have mated many hundreds of queens by this method.

DRONES

One further great advantage of this way of collecting up nucs was that you got very few drones with your bees, so you were not bringing home a whole lot of doubtful fathers to fly round your mating apiary. If you make up nucs with combs of brood, you will always get a little drone brood on the edge of these combs, unless you are very careful. This will mean more unknown parentage and also there will be drones on the combs themselves. It is always a good idea to minimise the chance of a poor mating. I can well remember my son going through nucs, frame by frame, and picking off the drones by hand, a time consuming operation!

Good drones are vital to a breeding venture and can be placed in your breeding apiary with careful planning. As I have said before, "Remember that drones do not have a father." This vital fact has great significance. If you bring daughters of the previous years breeding into your mating area then all the drones will be pure bred back to their grandmother. However, as she was the breeding mother of the previous year, you stand a very good chance of perpetuating good characteristics in your new stock, when it comes to mate. A planned breeding programme should always take into account this important fact. The more you can get the right fathers as well as the right mothers the better.

There is no doubt in my mind, that black drones will always fly at lower temperatures than will yellow drones. For this reason it will always be easier to mate a yellow queen with a black drone and, if I am right, this will explain why bees, left to their own devices, will inevitably get darker and darker. David Rowse pointed this out to me many years ago as something he had discovered and I confirm the fact.

If you include this thinking in your planning, it is quite clear that the more you err towards a yellow breeder queen the better, but do not overlook all the other far more important factors given earlier. Yellow queens are far easier to find in the hive so this will always be useful. If you breed from a black queen, I am certain all the daughters will be black and unless it is part of a plan to produce black bees, I would avoid it.

MATING

The next stage of the operation is the final one that of finding your beautiful young mated Queen laying her first eggs in the mini nuc. This will take place about ten days after she has hatched. If the weather is very good, that is, hot and fine, she may mate a little earlier, on the other hand if the weather is cold and wet, she may take up to two weeks. If it takes longer than two weeks I would become more and more doubtful about her future performance.

Sometimes a queen will mate after three weeks, but in my opinion that is the absolute limit and I would not be happy with such a queen. There are all kinds of problems that may result, such as poor mating, early drone laying or she may just never lay at all. I expect to see eggs on day ten, or maybe day eleven or twelve. If you have a batch of cells, all drawn at the same time, from the same breeder mother, then I would expect that they should all mate at around the same time. So once you see eggs in one, the others should follow very soon. Any queen that hangs around not mating, I would discard, or you may ruin a fine stock if you use her.

Bees will sometimes refuse to accept a new queen and I have known them to take two weeks to get a new mated queen back on to the lay, but this is unusual. Once a new queen has been caged into a stock I would leave her for a week. Hopefully by then young brood or at least eggs will be seen.

Queens can be taken from a nuc and used as soon as they are laying, but this may be over hasty. I have found that usually one in ten will turn out to be a drone layer and if you keep them in the nuc until you see sealed brood, you can then be more certain. Drone capped brood will appear at this stage and if it does she is obviously no good and should be killed. Also any queen seen laying more than one egg per cell is to be avoided.

In conclusion I can only say that the above is my method. There are many other ways of raising queens, but for me this method has worked. Furthermore, it is based on the work of David Rowse and Harry Wickens, both of whom produced many hundreds of fine queens over dozens of years and taught me my trade. Finally, the more we all work at our breeding the better will be our overall stock. With the advent of Varroa and the disappearance of almost all wild stocks of bees, the chance of our beautiful virgin queen mating with some rough local drone from a nearby church tower will soon cease to be a problem. The future must be better queens all round and in my opinion keep your affections for some good-looking local lass.

CURRENT REFLECTIONS

It is now ten years since I gave this talk on queen rearing and, having re-read my work, I am still quite happy with my approach to this subject. On the other hand since the advent of Varroa, there have been fundamental changes to our thinking about our bees.

Varroa has brought with it many unforeseen problems for the honey farmer. It would appear that it comes with a number of in-built diseases, which will spread through our colony from bee to bee and hive to hive. Many of these ailments may cause the colony to collapse and die. Of one thing I am sure, it is not the bees' fault. It is, as usual, the stupidity of man that has caused the problem.

When I first took up bee farming, some forty years ago we faced a few major diseases, especially foul brood of both types and a bit of Nosema. We struggled and we treated and we got by, but once Varroa got established, things went from bad to worse, if you did not look after your bees, you lost them, it was as simple as that.

It is my opinion that part of the trouble was the emergence and gathering pace of the importation of queens from various parts of the world. It was this influx of foreign bees that set us off down the slippery slope to where we are today. In *Honey by the Ton* (Barn Owl Books, London, 1983) I dealt quite fully with this question of queens from abroad, for in those far off days I tried them all and as I have said I never found any that were any good.

It was for this reason that I abandoned the practice and decided that the only queens that were going to be any good were those that I bred myself. I stuck by this decision and it is the policy that my son Robert still follows today, now that he has the reins and I am just a bystander.

Many people continued to import queens and sell them into the beekeeping world of the UK. Very slowly the British original strains began to disappear and people bought more and more queens from America, Australia, New Zealand and Jamaica. Very few people, outside UK honey farmers, even bothered to breed any indigenous stock and as fast as their new, and quite unsuitable, queens died off they went and bought some more. This created a lovely trade for the importers, but did irreparable damage to our native bees leaving a few safe pockets of breeding stock in England and a very safe Scotland!

At this point we must consider what was happening to the vast majority of our beekeeping. As Varroa extended its range and spread into all the feral colonies around the country, it killed all those that were not treated which would have been almost all of them.

Of course, this meant that there were fewer and fewer places where young queens could get mated by an indigenous drone. There just were not any about.

This was a disaster for almost everybody. We were flooding our countryside with a huge mob of quite unsuitable drones. As there were now far fewer bees of British original stock about; all our new young queens were getting mated by foreign drones, which was far from ideal.

At Field Honey Farms we have kept a breeding programme going and most of the queens that we use are home bred. We now keep a healthy population of indigenous bees on Salisbury Plain. This is an area where for very many years John Rawson ran his bee farm and we took over his sites and bees, when he retired. This has meant that a good healthy stock of drones has always flown in that part of the world, nearly all of them original local stock going back years. Such bees might be a little sharp tempered and inclined to swarm early, but they were British through and through!

These bees have evolved to fit in with the local flowering patterns and this is most important. They will build up in the spring on the golden fields of wild dandelion and will reach their first swarming phase early. There will then be plenty of good quality drones flying, which will pull themselves back together on the late May honey flow from sainfoin. They will build up on the summer clovers, vetches and trefoils, such as Yellow Melilot the queen of honey plants. The stocks will then give a good crop of honey, right into August. This is a strain that will fit in with nature, unlike some New Zealand stock that I have seen, which had drones flying in November and December. I saw them and asked myself "What sort of future have they got?"

One reads about drone collecting areas where queens go to mate and I am hopeful that there are one or two good places left on The Plain where our queens can congregate and meet with a better class of drones!

I am quite sure that this question of drone collecting areas is of great importance and it could be that without them, our poor quality imported stock, has nowhere to go, when their daughters come out to mate. It is obvious to me that indigenous stock will develop a mating pattern over many decades, if not longer. If these patterns are interfered with then the whole population will suffer. Drones will not meet queens, queens will not meet drones, and ancient collecting areas will disappear. I believe that we are in danger of going on down this route until there are no original indigenous bees left.

I give one example, with which I am familiar. For many years now we have stocked our rivers with brown trout brought up in stew ponds and they are quite unable to follow a pattern of breeding in the right place and at the right time. This will mean that eventually there will be no trout to catch in our rivers.

I can only hope that we do not do the same thing to our bee population. Breeding good queens is a time consuming process and I should know, for I have been trying to do that for over thirty years. Now I can only watch what others are doing, cross my fingers and hope.

One change that my son has made, which makes a huge difference to his breeding programme and the way he carries out his queen rearing and his re-queening is the use of a mobile warming box. (see illustration on p.20) This box is insulated, has a thermostatic control and will carry about twenty ripe queen cells. It can be carried in our vehicle and plugged into the lighter socket. With this box it is now possible to take a queen cell that is about to hatch and place it in a hive that needs a new queen.

The receiving hive will be one that has just lost its queen or had her removed because she is old, failing or maybe vicious. Often a ripe queen cell will be better received than a new young queen, as she may not be accepted. This is especially true if the bees are being kept far from home and travelling costs have to be taken into account. In this case, as long as the virgin mates alright, a fine new queen will be found on the next visit. But as always with bee farming, mistakes will be made and these will have to be corrected.

This also very much reduces the amount of queen rearing work that has to be done at home and the cells can be distributed during the usual day's work, thus solving problems as you go along. David Rowse often said to me that time and motion study was very important and that one should never work along a fixed line of ideas and not be prepared to change. With the current direction of fuel costs, there will have to be a great deal more planning on a bee farm. Gone are the days when R.O.B. Manley and Harry Wickens went by horse and cart from site to site at a time when they could not get the fuel because of the war.

Oliver S Field
September 2008

BEE BREEDING CHARTS

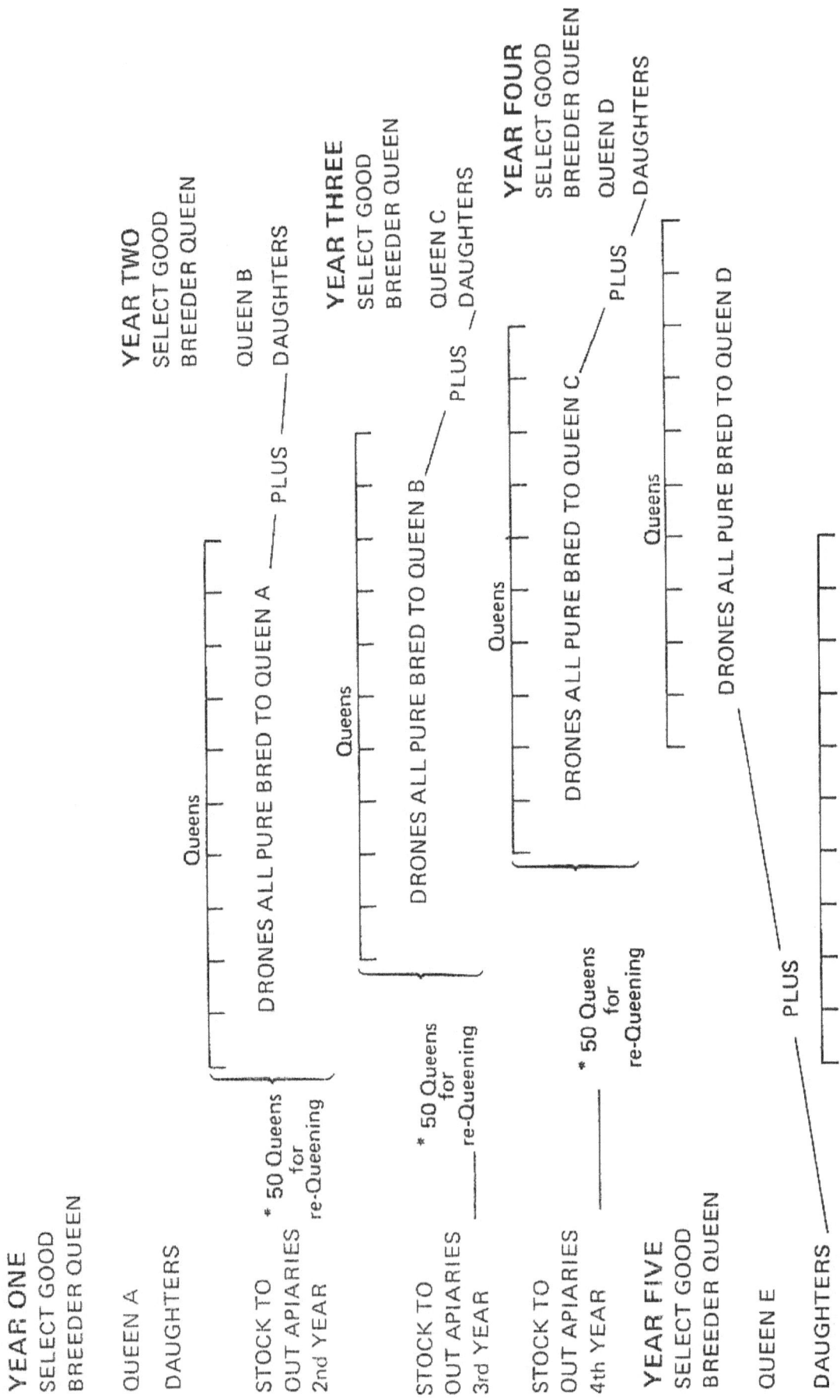

YEAR ONE
SELECT GOOD
BREEDER QUEEN

QUEEN A

DAUGHTERS

STOCK TO
OUT APIARIES
2nd YEAR

* 50 Queens
for
re-Queening

Queens

DRONES ALL PURE BRED TO QUEEN A

PLUS

YEAR TWO
SELECT GOOD
BREEDER QUEEN

QUEEN B

DAUGHTERS

STOCK TO
OUT APIARIES
3rd YEAR

* 50 Queens
for
re-Queening

Queens

DRONES ALL PURE BRED TO QUEEN B

PLUS

YEAR THREE
SELECT GOOD
BREEDER QUEEN

QUEEN C

DAUGHTERS

STOCK TO
OUT APIARIES
4th YEAR

* 50 Queens
for
re-Queening

Queens

DRONES ALL PURE BRED TO QUEEN C

PLUS

YEAR FOUR
SELECT GOOD
BREEDER QUEEN

QUEEN D

DAUGHTERS

Queens

DRONES ALL PURE BRED TO QUEEN D

PLUS

YEAR FIVE
SELECT GOOD
BREEDER QUEEN

QUEEN E

DAUGHTERS

YEAR ONE
BEE FARM
250 STOCKS

SELECTED BREEDER
A

YEAR TWO
250 STOCKS. RE-QUEEN
50 A STRAIN

SELECTED BREEDER
B

YEAR THREE
250 STOCKS. RE-QUEEN
50 B STRAIN

SELECTED BREEDER
C

YEAR FOUR
250 STOCKS. RE-QUEEN
50 C STRAIN

SELECTED BREEDER
D

YEAR FIVE
250 STOCKS. RE-QUEEN
50 D STRAIN

MUCH IMPROVED STOCK SHOULD NOW
BE AVAILABLE

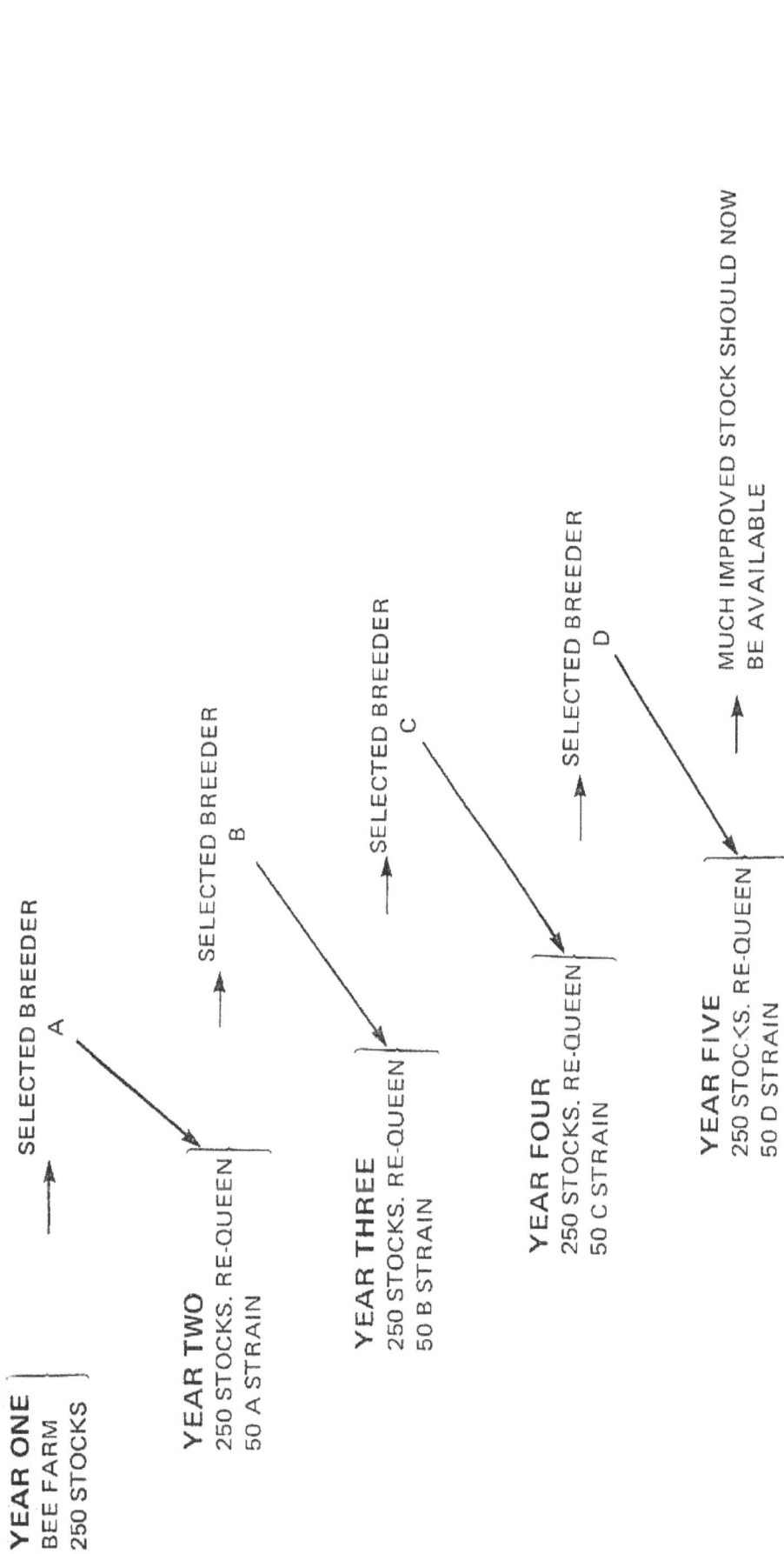

Breeding chart. The objective is to maintain pure-bred drones from one year's breeder queen to mate with daughters of the next year's breeder queen.